STEM in the MILITARY

ROCKETS and SATELLITES in Warfare

By Tanner Billings

Cavendish Square

Published in 2023 by Cavendish Square Publishing, LLC
2544 Clinton Street, Buffalo, NY 14224

Website: cavendishsq.com

Cataloging-in-Publication Data

Names:Billings, Tanner.
Title: Rockets and satellites in warfare / Tanner Billings.
Description: New York : Cavendish Square, 2023. | Series: STEM in the Military| Includes glossary and index.
Identifiers: ISBN 9781502665546 (pbk.) | ISBN 9781502665553 (library bound) | ISBN 9781502665560 (ebook)
Subjects: LCSH: Astronautics, Military --Juvenile literature. | Cold War --Juvenile literature. | Space surveillance --Juvenile literature.
Classification: LCC UG1520 B55 2023 | DDC 623.74/9--dc23

Editor: Kate Mikoley
Copyeditor: Abby Young and Michele Suchomel-Casey
Designer: Andrea Davison-Bartolotta

The photographs in this book are used by permission and through the courtesy of: Cover Brent Wong/Shutterstock.com; p. 5 Alones/Shutterstock.com; p. 6 wikimedia commons/File:GPS Blog IIIA.jpg; pp. 8, 26 U.S. Air Force photo; p. 9 United Launch Alliance/U.S. Air Force; p. 11 (top) Bill Morson/Shutterstock.com; p. 11 (bottom) Claudine Van Massenhove/Shutterstock.com; p. 14 betto rodrigues/Shutterstock.com; p. 16 courtesy of the Library of Congress; p. 17 U.S. Air Force photo/ Senior Airman Bobby Cummings; p. 20 (both) courtesy of NASA; p. 27 courtesy of U.S. Army; p. 28 wikimedia commons/File:Explorer 1 During the Installation to Jupiter-C.jpg; p. 29 Asetta/ Shutterstock.com; p. 32 wikimedia commons/File:Galactic Radiation and Background satellite 1.jpg; p. 34 Everett Collection/Shutterstock.com; p. 36 Rainer Lesniewski/Shutterstock.com; p. 39 wikimedia commons/File:STS-44 DSP deployment.jpg; p. 40 courtesy of Lockheed Martin Space Systems/U.S. Air Force.

CPSIA compliance information: Batch #CWCSQ23: For further information, contact Cavendish Square Publishing LLC at 1-877-980-4450.

Printed in the United States of America

CONTENTS

INTRODUCTION

The word "rocket" can be used to describe many different things. The term is often used generally to refer to various kinds of flying weapons. Missiles and bombs are sometimes described as rockets. Some types of fireworks are called rockets too. If you're moving really fast, someone might even call you a rocket!

There are several different types of rockets, and they have many different uses. A rocket engine is a jet engine powered by gases released from burning fuel. Sometimes the word "rocket" refers to the engine itself. Other times it refers to the vehicle it is powering, such as a spacecraft, missile, or other weapon. In very general terms, a rocket is a vehicle that can launch into the sky.

Rockets are what have made space exploration possible. They're especially useful because they can operate both in space and within Earth's atmosphere, meaning they can be used for takeoff from Earth but then still continue to work when they cross into space. So far, rockets are the only type of engine we have that can do that.

The U.S. military has many uses for rockets, both in war and peacetime. The Air Force Space Command (AFSC) was established in 1982 to handle military-focused space systems.

When you think of a rocket taking off, you might picture something like this. The flames and smoke are a result of the rocket's fuel.

Then, on December 20, 2019, the U.S. Space Force (USSF) was established. The AFSC was redesignated as the USSF. This meant people who had worked for the AFSC were now assigned to the USSF.

As of 2022, the USSF is the newest branch of the U.S. military. Its job is to organize, train, and equip forces for space. This means part of the job includes dealing with rockets and satellites. While we hope to never have to fight a war in space, the USSF is prepared to take one on if the time ever comes.

Satellites, such as this Global Positioning System (GPS) satellite, are launched into space with rockets.

CHAPTER 1

THE BATTLE FOR SPACE

An artificial satellite is a machine sent to space to orbit a planet, the sun, or the moon. It might not seem like it, but satellites are an **integral** part of life on Earth. Satellites observe and study the objects they orbit, and they're also used for important communication systems. Radio, television, and internet systems often depend on satellites.

If you've ever used your phone to figure out how to get somewhere, chances are you were using satellite technology. GPS, or the Global Positioning System, is a satellite navigation system operated by the USSF. The satellites that are part of this system are what help you find directions when you use navigation apps such as Google Maps or Apple Maps.

The USSF is charged with, among other things, protecting U.S. satellites from attack. This is an important duty, because without these satellites, life as we know it would be very different.

Top-Secret Space Plane

One possible means of defending U.S. satellites and interests in space is the super-secret Boeing X-37B space plane. This space plane, also known as an Orbital Test Vehicle (OTV), is a space vehicle that can be returned to Earth and used again. However, due to its top-secret nature, some observers think it may be used for other purposes.

Beginning with the launch of OTV-1 from Cape Canaveral, Florida, on April 22, 2010, the X-37B has successfully flown six missions as of 2022. The air force revealed very little about OTV-1's mission, but it was placed in an orbit typically used for military surveillance satellites. It landed at what was then known as Vandenberg Air Force Base in California 224 days after launch.

The sixth launch for the X-37B happened in May 2020. The OTV is an unmanned **reentry spacecraft** and looks very much like a mini-space shuttle.

The U.S. military defines the X-37B Orbital Test Vehicle as part of "an experimental test program to demonstrate technologies for a reliable, reusable, unmanned space test platform." It is designed to accomplish two things: to provide a reusable vehicle to advance the United States' future in space and to facilitate experiments in space that can be returned to Earth and evaluated.

While the X-37B is in orbit, it is controlled like any other satellite. Operators on the ground monitor its **telemetry** and send signals to maintain its assigned flight plan. At the end of its scheduled flight, an operator signals a command to the spacecraft to return to Earth. This triggers the vehicle's orbital maneuver engine. The X-37B then descends on its own through the atmosphere and lands by itself on a specified runway. It is flown by an onboard computer.

The X-37B is launched with the help of a very powerful rocket. After liftoff, the space plane is released into space for its mission before it returns to touch down on Earth, landing as a plane.

Joint Space Operations Center

The main space command site is the Joint Space Operations Center (JSPOC) at Vandenberg Space Force Base in California. The JSPOC was created in 2015 to provide a single location for tracking satellites of both intelligence agencies and the Department of Defense (DoD). The opening of the JSPOC allowed for intelligence and military agencies to work together in developing tactics, techniques, and procedures for fighting a war in space. When it opened, Vandenberg was still an air force base. The air force and the National **Reconnaissance** Office (NRO) worked together to develop the center. The NRO controls the construction and operation of the nation's satellites.

The command center was created to meet increased threats to U.S. interests in space. Notable threats at the time included China's antisatellite (ASAT) missile test and Russia's **annexation** of Crimea and attempts to undermine Ukraine's government. American observers viewed Russia's growing hostility as a danger both to the world and to the boundless limits of space. In fact, after Russia invaded Ukraine in 2022, the USSF observed that Russian forces attempted to jam GPS signals used by Ukrainians. While there was no sign that Russians had directly attacked any satellites, the USSF was carefully on guard.

Vandenberg Air Force Base was renamed Vandenberg Space Force Base on May 14, 2021.

VANDENBERG
SPACE FORCE BASE

VANDENBERG
SPACE FORCE BASE

Other OTV Missions

The OTV-2 mission, using a second X-37B, launched from Cape Canaveral on March 5, 2011. The air force announced it would extend the vehicle's in-orbit baseline time of 270 days but described its mission only as a test of new space technologies. OTV-2 completed a successful mission and landed 469 days later at Vandenberg.

On December 11, 2012, OTV-3, using the first X-37B, launched from Cape Canaveral and further extended the vehicle's in-orbit time. It landed successfully at Vandenberg after 674 days in orbit on an undisclosed mission.

A fourth mystery mission of the X-37B (OTV-4) began when it shot into space atop an Atlas V rocket on May 20, 2015. Its announced mission was to conduct scientific experiments. On September 7, 2017, OTV-5 was launched. Both OTV-4 and OTV-5 landed at Kennedy Space Center in Florida.

Spies in the Sky

In modern warfare, knowing the enemy is essential to the success of any military operation. The use of spies on the ground to gather information on the enemy dates back to the time of humankind's earliest armed conflicts. Both the Union and Confederate armies used hot-air balloons for reconnaissance during the American Civil War (1861–1865). In World War I (1914–1918), all sides continued to use balloons and added fixed-wing aircraft to observe enemy operations. During World War II (1939–1945), high-flying aircraft began taking high-**resolution** photographs to gather information.

After World War II, reconnaissance aircraft flew higher yet. However, even at great heights, they became vulnerable as the world entered the high-tech age. Technology has moved the art of **espionage** to the high frontiers of space.

Today, many military analysts suspect the X-37B is up to a lot more in space than simply scientific experiments. Some think that it may be used for reconnaissance activities or spying.

Regardless of what exactly the space planes are doing, one thing is for sure: Satellites act as our eyes in the sky. They keep the nation informed as to what its potential enemies are doing or not doing. They also keep many of our communication systems running. The X-37B orbiter may play a key role in defending American satellites in space—both now and in the future.

In order to fully understand why satellites are so important to the world as we know it today, we have to go back in time. By seeing what certain tasks were like before satellites, we can better appreciate their benefits.

Join the Debate

1. How important do you think satellites are to everyday life in the United States?
2. Why do you think the nature of X-37B missions is kept secret?
3. Do you think the Space Force needed to become its own branch in the military? Why or why not?

The Lockheed U-2 was designed by the renowned aircraft designer Clarence "Kelly" Johnson in 1953. The U-2S, shown here, is a more advanced version of the original.

CHAPTER 2

AN UNFORTUNATE U-2 INCIDENT

In Earth's atmosphere, high-flying reconnaissance flights pose a serious risk to both the plane and the pilot. That's one reason why some people think reconnaissance can be more safely accomplished with unmanned satellites in space. Still, manned planes have long proven useful for spying on the enemy. The Lockheed U-2 is among the best. It is a single-engine, sailplane-winged spy plane developed for the U.S. Air Force and the Central Intelligence Agency (CIA). Nicknamed the "Dragon Lady," it entered service in 1957. The last U-2 was built in 1989, and the planes are still in service today.

The technology used by the Lockheed U-2 is very advanced, and the aircraft have proven to be incredibly useful. Still, one major incident is probably responsible for the U-2's reputation as possibly the most famous spy plane in the world.

The Incident

An event now known as the U-2 incident took place on May 1, 1960. It began when CIA pilot Francis Gary Powers climbed into the cockpit of his U-2 spy plane in Pakistan. The high-altitude aircraft flew at a maximum height of 70,000 feet

(21,336 meters). Moments later, Powers took off from a U.S. base at Peshawar and set out on a mission code-named Grand Slam.

His flight was to take him around 2,900 miles (4,667 kilometers) across the **Union of Soviet Socialist Republics** (USSR) to a destination at the top of the world in Norway. The chief aim of his mission was to fly over and photograph Soviet **intercontinental ballistic missile** (ICBM) sites at Baikonur and Plesetsk.

It is illegal for aircraft of one nation to fly over another nation without permission, but Powers could take comfort in the knowledge that no Soviet aircraft or missile could reach him at 70,000 feet (21,336 m). He had every reason to feel secure and out of harm's way. After all, the United States had been conducting reconnaissance flights safely over the Soviet Union since 1956. Although the Soviets knew about the U-2 overflights, they were powerless to stop them—until the day of Powers's flight.

Suddenly, high above the Soviet city of Sverdlovsk Oblast in the Ural Mountains, an SA-2 surface-to-air missile caught up with the U-2 and shattered the high-altitude quiet and sanctity of its American pilot. Powers ejected at about 34,000 feet (10,363 m) and parachuted to the ground. He carried a toxic needle with him to choose to die rather than reveal top-secret information to his captors, but he did not use it. Upon landing, he was captured by the KGB (Soviet security agency), tried for espionage, and imprisoned.

Powers, who is shown here, died in a helicopter crash in August 1978. He was working as a pilot for a Los Angeles television station.

Merryl Tengesdal

Merryl Tengesdal had a dream. She imagined herself soaring above Earth into the deepest reaches of space. However, she was no idle dreamer. She acted on her dream and became a colonel in the U.S. Air Force and one of only a handful of female pilots who fly the U-2 reconnaissance aircraft. She is the first, and so far only, African American female U-2 pilot in history.

"The Air Force has always been on the forefront of breaking aviation and racial barriers," Tengesdal said. "I am extremely proud of being the first Black female U-2 pilot in history."

Things did not come easy for Tengesdal. After receiving a degree in electrical engineering, she graduated from the U.S. Navy's Officer Candidate School as an ensign in 1994. She attended flight school and earned her wings. Ten years later, she transferred her commission to the air force and successfully completed the demanding U-2 pilot training.

Tengesdal has flown multiple missions over Iraq and Afghanistan and elsewhere in support of U.S. troops. "From my time in the navy to my experiences in the U-2 program," she said, "I like to think I've played a part in helping some of the troops on the ground get home safely."

Tengesdal also served as director of inspection for the Air Force Inspector General. Now retired, she is a motivational speaker who talks about overcoming adversity and encourages children of color to become involved in science, technology, engineering, and math (STEM) fields.

After serving 18 months of a 10-year sentence in a prison east of Moscow, Powers was released in exchange for a Soviet spy known as Colonel Rudolf Abel.

Going Down

American officials later learned that the Soviets had fired a volley of SA-2 missiles at the U-2. The Soviets also shot down at least one of their own MiG jet fighters that had been pursuing Powers. It remains unclear exactly how the U-2 was destroyed. The SA-2 missiles should not have been able to reach the spy plane at that altitude. One explanation suggests that the fragile aircraft might have been sufficiently damaged by the shock waves of the exploding missiles.

The U-2 incident could not have happened at a worse time for President Dwight D. Eisenhower and the United States. Soon after World War II ended, the United States entered into an arms race with its former ally, the USSR. Each nation tried to outdo the other with more and better weaponry. The Soviets tested their first atomic bomb in August 1949. For the next four decades, a period of open, but restricted, rivalry and hostility existed between the two nations and their respective allies. The period came to be known as the Cold War.

By 1955, officials in both nations had grown deeply concerned about the nuclear capabilities of the other. President Eisenhower proposed an "open skies" plan at a conference in Geneva, Switzerland. His plan would allow each country to make overflights of the other for conducting mutual aerial inspections of nuclear facilities and launch pads to assure both nations of their shared interest in preventing a nuclear firestorm.

Soviet premier Nikita Khrushchev rejected Eisenhower's proposal. His nation, he said, held to an established policy of not allowing international inspections of any kind. U.S. officials scrambled to find new ways of verifying the state of nuclear development in the USSR. Eisenhower was determined to reach a nuclear arms agreement with the Soviets. He arranged a four-power peace summit in Paris, France, between the United States, the USSR, Britain, and France in mid-May 1960.

In the meantime, news broke about the Soviet downing of a U.S. spy plane. Unaware that Powers had been captured, U.S. officials issued a series of cover stories denying that Powers was on a spy mission. The Soviets quickly exposed the stories as lies. Eisenhower admitted the spy flight and later efforts to deny it—but the damage had been done.

What Would Be Next?

After the U-2 incident, Eisenhower felt so distressed, he reportedly told his secretary about his desire to resign. In the increasingly hot environment of the Cold War, it became clear that the United States needed a new way to keep an eye on its enemies.

It wouldn't be long until a new method of reconnaissance was born. With the advent of rockets and artificial satellites, spy missions would become much harder for enemies to detect—and even if they could detect them, they would hopefully be much safer for the humans operating them.

Join the Debate

1. Do you think anything could have been done to avoid the U-2 incident?
2. Why do you think President Eisenhower wanted to resign after the U-2 incident?
3. In what ways do you think satellites may be safer than spy planes?

John Glenn

In the continuing race for space, astronaut John Glenn was launched into space on *Friendship 7*, in 1962. It was the first American manned orbital space flight. The rocket was a modified Atlas intercontinental ballistic missile.

CHAPTER 3

ARTIFICIAL SATELLITES AND ROCKETS

The time after World War II saw not only an arms race between the United States and the USSR. It also saw a space race. Scientists and engineers from both countries worked tirelessly trying to develop rockets strong enough to make it to space. Both countries wanted to beat the other to it.

While the technology those involved in the space race were developing was new, rockets were not. In fact, the first rocket dates all the way back to ninth-century China, after the discovery of gunpowder. It was a firework missile. As weapons, rockets were used by Chinese armies as early as 1232 CE.

Rockets in War

The military use of rockets quickly spread throughout Asia and Europe. These early rockets burned a substance called black powder as a **propellant**. It consisted of charcoal, saltpeter, and sulfur. British soldiers used rockets against Americans in the War of 1812. Americans used rockets in the war against Mexico (1846–1848). During the American Civil War, both sides used rockets. The French shot down enemy observation balloons with rockets in World War I. Germany launched

V-2 rockets against Britain during World War II. The V-2 was developed by a team of German scientists headed by Wernher von Braun and others. After the war, von Braun and about 150 other German scientists helped to develop rockets used in the United States' entry into the space age.

Robert H. Goddard

In America, physicist and inventor Robert H. Goddard (1882–1945) is regarded as the father of modern rocketry. He was born to a middle-class family in Worcester, Massachusetts, at the height of the post–Civil War Industrial Revolution. Influenced by H. G. Wells's space novel *War of the Worlds*, he began dreaming about space flight at an early age.

In 1926, Goddard conducted the first successful launch of a rocket using liquid fuel at his aunt's farm near Auburn, Massachusetts. He went on to launch larger rockets in the 1930s. His contributions to rocket science resembled a shopping list. Notably, he proved that a rocket would work in a vacuum (such as space) and needed no air to push against. He used vanes on a rocket motor for guidance, and he developed a gyroscopic control apparatus for directing a rocket. Goddard's illustrious career was cut short in 1945 when he died of throat cancer.

All About Rockets

So, what exactly is a rocket? It is a type of engine that produces **thrust** to push itself upward or forward. A rocket is also the name for a vehicle or an object that is driven by a rocket engine. An engine produces thrust by expelling gas. The massive force (or action) generated by hot gases firing out the back of a rocket engine creates an equal force (or reaction) that pushes the rocket vehicle forward in space. (This is an example of Newton's third law of motion: For every action there is an equal and opposite reaction.) Unlike combustion engines, rocket engines do not require outside air (oxygen) to function, thus they can operate in space where combustion engines are inoperable.

The basic parts of a rocket are the propellants, chamber, nozzle, and payload. The chamber is the section of the rocket where propellants are put under pressure. The nozzle is the opening at the end of the chamber through which the pressurized propellant is released. The payload is the cargo—such as an artificial satellite—carried in the nose cone at the top of the rocket.

When it comes to propellants, there are three main types: liquid, solid, and hybrid. Almost all rocket propellants consist of an **oxidizer** and a fuel. Today's standard high-energy solid rocket fuel consists of a mixture of ammonium perchlorate powder (oxidizer) and fine aluminum powder (fuel). They are held together by a rubbery base mixture. Liquid oxygen (LOX; an oxidizer) and liquid hydrogen (fuel) are the commonly used liquid propellants in the latest U.S. rockets. Hybrid propellants combine a solid fuel and a liquid or gas oxidizer.

Artificial Satellites

In contrast to the centuries-old concept of rockets, the first artificial satellite came from a union of contemporary science and technology.

An artificial satellite—as opposed to a natural satellite like the moon—is a man-made object that continuously orbits Earth or another spatial body. Most artificial satellites orbit Earth, but some orbit the sun, the moon, asteroids, and planets, usually to gather information. Orbiting spacecraft—space capsules, space shuttle orbiters, and space stations—are also considered satellites.

How does a satellite stay in orbit? The key is a balance between its velocity (the speed at which it would travel along a straight line) and the gravitational force between the satellite and Earth. Without the pull of gravity, the satellite's velocity would propel it into space in a straight line. Conversely, without velocity, gravity would pull the satellite back to Earth.

According to their mission, artificial satellites are classified into six main types: scientific research, weather, communications (sometimes known as SATCOM), navigation, Earth-observing, and military. Spy satellites are grouped in the latter category and are generally launched in a 90-minute polar orbit.

Satellite Orbits

There is an infinite variety of satellite orbits. Four common orbits are a high-altitude geosynchronous orbit, a medium Earth orbit, a sun-synchronous polar orbit, and a low Earth orbit.

A high-altitude geosynchronous orbit is positioned above the equator about 22,000 miles (35,400 km) high. It circles Earth's axis in the same direction and time as Earth rotates on its axis. Because it remains "in sync" with Earth's rotation, it appears to be stationary, hence its orbit is also known as geostationary.

A medium Earth orbit circles Earth in 12 hours at an altitude of about 12,400 miles (20,000 km).

A sun-synchronous polar orbit maintains a constant relation to the sun—it orbits at the same speed Earth is turning and seems to stay in place—at a fairly low altitude and passes almost directly over the North and South Poles.

A low Earth orbit is positioned just above Earth's atmosphere, where there is almost no air to produce drag on a spacecraft and slow its speed. Less energy is required to put a satellite in this orbit.

SATCOM Operators

Communication is vital to an army in the field—even more so in today's high-tech environment. That is why satellite systems operators and maintainers are so important to the U.S. Army. SATCOM operators are responsible for keeping lines of communication always up and running. They operate and maintain multichannel satellite communications for the entire army.

To qualify as a SATCOM operator, a candidate must first take the Armed Services Vocational Aptitude Battery. This is a series of tests that helps the army determine what jobs are best suited to the applicant.

After 10 weeks of basic combat training, the candidate receives 15 weeks of advanced individual training with on-the-job instruction. SATCOM training includes time spent in both the classroom and in the field.

Candidates learn how to use various types of communication equipment, how to work with codes, and how to maintain and care for communications equipment. SATCOM operators span the globe and extend the reach and awareness of the U.S. Army everywhere.

The Wideband Global SATCOM satellite is a communication satellite that increases the communications capabilities of the U.S., Canadian, and Australian militaries.

Soldiers in the U.S. Army use satellite dishes and other communication equipment to stay connected.

For nearly any military job working with satellites or rockets, an important trait is a thorough knowledge of STEM skills. STEM stands for science, technology, engineering, and math. SATCOM operators need to be well-versed in these subjects. If you think you might like to work with satellites in the military someday, it's a good idea to start studying these subjects now.

Join the Debate

1. Why do you think the space race was important to American and Soviet citizens alike?
2. What do you think is the most important responsibility of a SATCOM operator?
3. Why do you think it's important for SATCOM operators to be well-versed in STEM subjects?

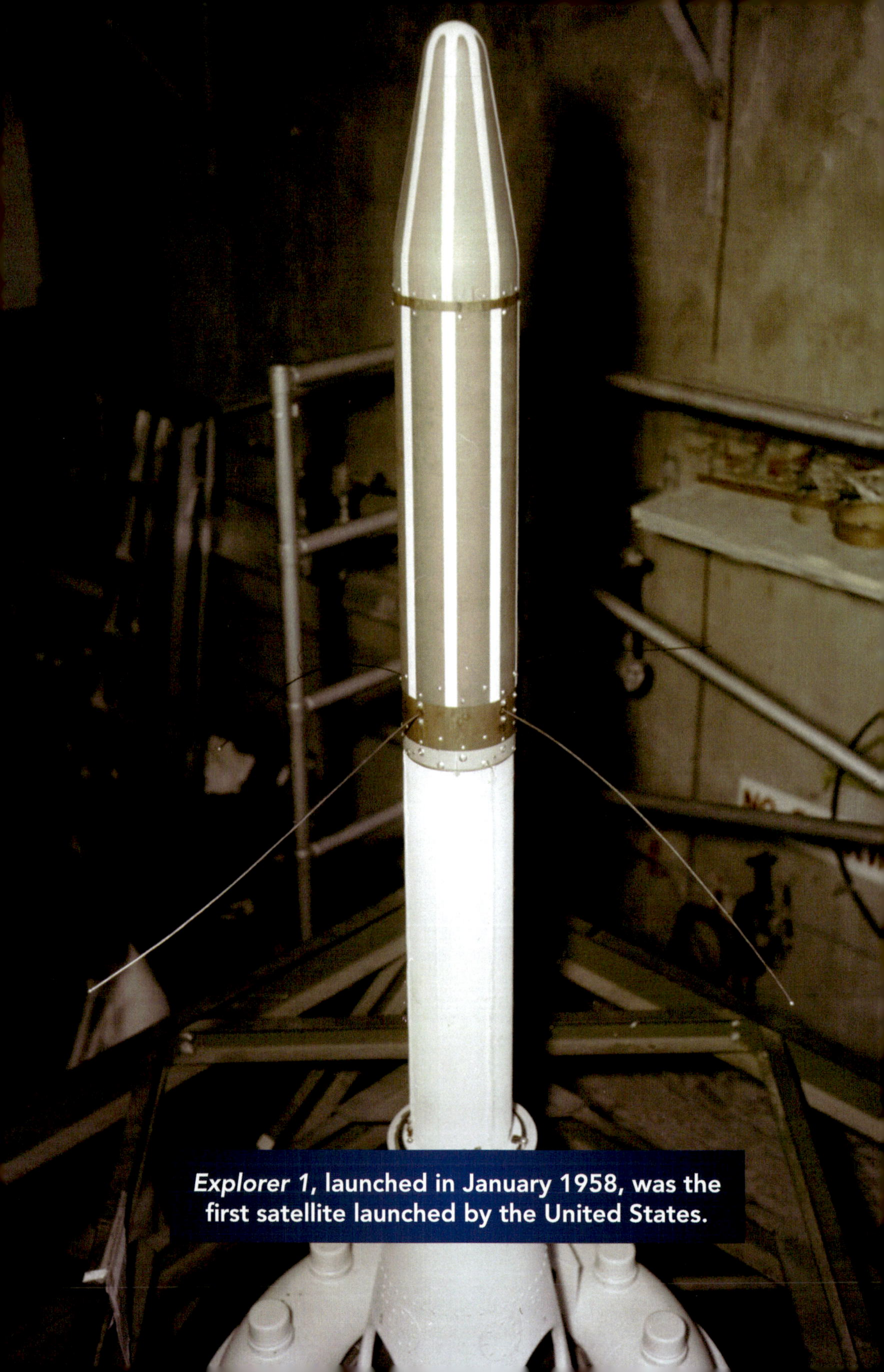

Explorer 1, launched in January 1958, was the first satellite launched by the United States.

CHAPTER 4

The SAMOS, MIDAS, and CORONA Programs

The first artificial satellite to orbit Earth was *Sputnik 1*, launched by the Soviets on October 4, 1957. The Soviets didn't stop there. Just a month later, on November 3, 1957, they launched a dog named Laika into space. When Laika, whose name was Russian for "Barker," went to space on *Sputnik 2*, she became the first living creature sent to orbit Earth.

After these Soviet achievements, Americans felt the need to catch up. In January 1958, the United States launched the *Explorer 1* satellite. Then, in March of the same year, the country launched the *Vanguard 1* satellite. While these were helpful, the United States was still largely dependent on reconnaissance data from the U-2.

Laika, sometimes referred to as the Space Dog, was the first living creature in orbit. Sadly, she died in flight.

Weapons System 117L

In July 1956, the U.S. Air Force approved Weapons System 117L

(WS-117L). It was a defense program to build reconnaissance satellites for use at varying orbit heights. By the end of 1959, WS-117L had evolved into three separate programs. They were the Discoverer program (code-named CORONA), the Satellite and Missile Observation System (SAMOS), and the Missile Defense Alarm System (MIDAS). Discoverer and SAMOS were designed for photo reconnaissance missions. MIDAS was created to provide early warning of hostile missile launches.

SAMOS

SAMOS carried a heavier satellite than the Discoverer. It was launched by an Atlas booster rather than the Thor used for the Discoverer. SAMOS was designed to take photographs of Earth, develop the pictures on board, scan them electronically, and then radio them to Earth. All this was to be done in near real time. If the satellite's electronics failed, it could rocket exposed film into space and parachute it to Earth in a canister. The first launch of SAMOS in October 1960 failed; a second in January 1961 succeeded. Thereafter, the SAMOS program mysteriously disappeared in a veil of secrecy. It has never been declassified.

MIDAS

The MIDAS program was a U.S. system of 12 early-warning satellites. They provided limited notice of Soviet ICBM launches between 1960 and 1966. The first MIDAS satellite was launched in February 1960 and failed to achieve orbit. The launch of *MIDAS 2* succeeded the following May, but its telemetry failed two days after launch. The launch of *MIDAS 3* achieved orbit in July 1961. It was launched by the Atlas-Agena rocket and was the heaviest U.S. satellite launched up to that time.

Much like SAMOS, MIDAS was soon shrouded in secrecy by the U.S. government. However, the program eventually led to the development, launch, and operation of the Defense Support Program (DSP). The DSP mission was declassified after the Gulf War (1990–1991). Today, DSP satellites, ground stations, and mobile ground terminals perform MIDAS's original mission of detecting and reporting on hostile missile launches.

While the DSP system was in development, CORONA cameras kept a photographic eye on the Soviet Union, China, and other potentially hostile nations.

CORONA

The CORONA program—named on a whim for a Smith-Corona typewriter—started under the name Discoverer but was renamed CORONA by the CIA. It was conceived in Sunnyvale, California, in 1958.

CORONA consisted of a series of reconnaissance satellites with KH designators (KH for "keyhole"). As a film-return program, it was intended to form a temporary bridge between U-2 reconnaissance flights and the development of more sophisticated air force spy satellites.

To avoid interservice rivalries, President Eisenhower directed the CIA and the air force to share joint responsibility for the program. The CIA managed the program, including secret financing; the air force provided the Thor-Agena rockets, launch pads, and recovery aircraft.

The CIA's history of the program, noted by William E. Burrows in *This New Ocean*, revealed their rocky working relations: "The vague assignments of responsibilities caused no appreciable difficulties in the early years … but they later (1963) became a source of severe friction between CIA and the Air Force over responsibility for running the program." To some, control of the program became a battle, but a battle well worth the struggle.

After 12 unsuccessful launches, the CORONA project got it right with its first successful launch from Vandenberg Air Force Base on August 18, 1960. Its first flight covered more than 1.65 million square miles (4.273 million sq km) of Soviet territory

The Galactic Radiation and Background Satellite

The Galactic Radiation and Background (GRAB) satellite was the first successful U.S. electronic intelligence (ELINT) satellite. Conceived in the late 1950s, *GRAB-1* was successfully launched at Cape Canaveral on June 22, 1960, 52 days after the U-2 incident. *GRAB-1* began transmitting space-intercepted electronic intelligence signals to stations on Earth.

In 1998, Keith Hall, then director of the National Reconnaissance Office, paid tribute to workers on the GRAB program: "Despite the setbacks [three out of five failed launches], they had the vision to see the possibilities of space, and the technical skill and tenacity necessary to break down the barriers to this new world of opportunity."

This program demonstrated that a space platform could collect as much ELINT as all sea-, air-, and land-based reconnaissance platforms operating within its field of view.

This model of a GRAB satellite is on display in the National Cryptologic Museum in Maryland.

and returned around 3,000 feet (914 m) of film. For the next 12 years, it continued to be America's eyes in the sky. Flights eventually lasted for as long as 19 days with multiple cameras capable of resolutions of 6 feet (1.8 m)—that is, able to distinguish objects 6 feet (1.8 m) wide from space.

CORONA Succeeds

CORONA satellites featured an evolving series of high-resolution cameras using Mylar film. Film was retrieved in recoverable film capsules, or "buckets," by aircraft or naval vessels. The 6593rd Test Squadron was formed to recover the film canisters, and a midair recovery system was developed, with a floatable canister as backup.

The film was exposed in space, transferred to a canister fitted with a parachute, and retro-rocketed into a reentry atmosphere. An airplane flying over the slowly descending parachute would then snag it with a specially designed hexagon hook.

The first attempt at a midair recovery failed, but squadron members managed to fish the film canister out of the ocean. A second attempt by the squadron in a C-119 "Flying Boxcar" succeeded on August 18, 1960.

Join the Debate

1. Do you think sending animals into space is ethical? Why or why not?
2. Why do you think it's important to have satellite photos of Earth?
3. Do you think it was a good or bad idea for President Eisenhower to have the CIA and air force work jointly on the CORONA program?

Apollo 11 was the spaceflight that brought the first humans to the moon. A Saturn V rocket launched it into flight.

CHAPTER 5

INTO THE MODERN AGE OF ROCKETS AND SATELLITES

The USSR continued to pull its weight in the fight to win the space race. On April 12, 1961, Soviet cosmonaut Yuri Gagarin rocketed into space on the *Vostok-1* spacecraft, becoming the first person in space. He completed a 108-minute orbital journey and returned safely to Earth. The Soviet achievement shocked U.S. space scientists and engineers. The Americans thought their technology was much more advanced than that of the Soviets. Americans thought they would put the first human in space. Gagarin's flight embarrassed them.

President John F. Kennedy, concerned over America's role in space, gathered his top advisers. According to Fordham space historian Asif Siddiqi, speaking with Mike Wall of SPACE.com, Kennedy asked, "What can we do to beat these guys?" Siddiqi went on. "They conferred for a few weeks, and they came up with the moon landing."

The rest needs little retelling. The United States won the race to the moon—but not without a nudge from the Soviets. This was a big achievement, but the study of space was far from over. Rockets and satellites still had a lot to accomplish.

The Search for bin Laden

On May 2, 2011, a team of U.S. Navy SEALs in helicopters dropped out of the sky in the dead of night in Pakistan and swept into the sprawling compound occupied by **al-Qaeda** leader Osama bin Laden. They left 40 minutes later with bin Laden's dead body. In addition, they carried a trove of computer devices and other intelligence information about al-Qaeda and the activities of its leader.

The search for bin Laden took months. Ultimately, the trail uncovered by human and satellite intelligence led to his walled compound in Abbottabad, Pakistan. The raid that killed bin Laden was directed by a fleet of satellites that collected electronic (ELINT) and signals (SIGINT) intelligence over Pakistan while the mission was happening.

This map shows the location of Abbottabad, Pakistan, where Osama bin Laden was killed on his compound.

Satellite technology provided U.S. officials in the White House with a front-row seat to the operation in Abbottabad in real time. John O. Brennan, then chief counterterrorism adviser to President Barack Obama, later said, “I’m not going to go into details about what type of visuals we had or what type of feeds that were there, but it gave us the ability to actually track it on an ongoing basis.”

History will record the demise of the terrorist leader as a triumph of American will, coupled with human and technological intelligence. Success of the operation owes much to the United States’ cutting-edge satellite reconnaissance systems that evolved over the last half-century.

Defense Support Program

The MIDAS satellite of the mid-1950s ultimately failed because its heat sensors kept picking up warm sunlight on cloud tops that triggered false alarms. However, it eventually matured into a long-lived class of early-warning satellites. The later offshoot system was named the Defense Support Program (DSP). Its satellites ring Earth in a geosynchronous orbit, 22,000 miles (35,405 km) above the equator.

DSP satellites use 12-foot-long (3.7 m) infrared telescopes to peer down from space. They rotate six times a minute to cover Europe and Asia with inquiring eyes. The elongated orbs continually search for rocket tests, space launches, and the blastoffs of multiple ballistic missiles that could signal the start of World War III.

Unlike its secretive MIDAS predecessor, DSP was highly publicized. American strategic planners wanted to make sure Moscow knew that the United States would see an attack coming and retaliate. The Soviets developed their own early-warning system, less reliable than the DSP satellites, but good enough.

Recognition of the ability of both sides to foresee an attack and strike back led directly to the doctrine of mutual assured destruction (MAD), which arose in the early 1960s. The thinking behind the MAD doctrine reasoned that neither side would launch a nuclear attack knowing the other side would immediately respond the same way.

The DSP is a system of satellites operated by the U.S. Air Force. As America's early-warning system, DSP satellites have been monitoring the skies for ballistic missile launches since 1970. Their ability to detect missile launches and provide data on such operations relies on two key factors: First, they must be able to find and identify the infrared radiation given off by missile trails; second, they must then be able to separate those emissions from various others caused by fires, explosions, and not least from Earth's natural glow.

DSP satellites form the main part of the Satellite Early Warning System (SEWS) used by the United States today. SEWS currently consists of five DSP satellites: Three provide frontline operational service, and two serve as available replacements should problems emerge with the primary satellites.

Titan IV and Delta IV rockets launched the DSP satellites into space. The last DSP satellite was launched in 2007. Since the program's inception, 23 satellites have been launched at an estimated cost of $400 million each. Additionally, improvements to the DSP via five upgrade sets have extended those satellites beyond their design lifespan.

This DSP satellite was deployed by Space Shuttle *Atlantis* in 1991.

Space-Based Infrared System

The DSP group is being replaced by the Space-Based Infrared System (SBIRS). While DSP satellites are still operational, the military has shifted its focus primarily to SBIRS.

The Space-Based Infrared System is a satellite system that provides data for missile surveillance, missile defense, technical intelligence, and battlespace awareness. It is designed to field a constellation of satellites in geosynchronous Earth orbit (GEO) and host payloads in highly elliptical orbit (HEO). An integrated centralized ground system under the U.S. Space Force will control the SBIRS constellation.

A Major Milestone

Early in 2015, airmen from the 460th Operations Group at Buckley Air Force Base, east of Denver, Colorado, established a major milestone in the Defense Support Program. Operators completed the first series of SBIRS satellite and antenna communications from Block 10, the new operations center for DSP satellites.

During an interview, SBIRS satellite engineering chief Captain Natasha Rosario told Airman 1st Class Emily E. Amyotte, "The intent of Block 10 is to bring all three platforms here in one place to the Mission Control Station." The first SBIRS command was sent on January 28, 2015.

Rosario, of the 2nd Space Warning Squadron, went on to say, "It was kind of fun. They sent the very first one, which was on DSP, and everyone clapped after the first command went out. They were like, 'Yes it worked!'" The milestone is one of more to come in a continuing effort to keep America's premier defense system second to none.

This SBIRS satellite is being worked on at Lockheed Martin Space Systems in California.

The Space-Based Infrared System is designed for rapid data collection and distribution. This is how it works: When a launch occurs, SBIRS sensors detect and track infrared radiation from the missile's hot exhaust. Data collected by onboard satellite systems is processed and transmitted to ground terminals around the world. The terminals forward the data to the Mission Control Station (MCS) at Buckley Space Force Base, Colorado.

TIMELINE OF ROCKETS AND SATELLITES

1232: Chinese armies use rockets for the first time.

1926: Robert H. Goddard conducts first successful launch of a liquid-fueled rocket.

1956: U.S. Air Force (USAF) approves WS-117L.

1957: Soviets launch *Sputnik 1*, the first artificial satellite to orbit Earth.

1960: Francis Gary Powers is shot down in a U-2 over the Soviet Union. CORONA satellite is launched successfully for the first time at Vandenberg Air Force Base.

1970: U.S. launches first DSP satellite.

1991: Soviet Union collapses.

2007: U.S. launches last DSP satellite.

2010: X-37B spacecraft flies first orbital mission.

2015: Joint Space Operations Center is created at Vandenberg Air Force Base. Boeing X-37B (OTV-4) is launched successfully into space atop an Atlas V rocket at Cape Canaveral.

2019: The Air Force Space Command is redesignated as the U.S. Space Force.

This timeline shows the progression of how rockets and satellites have been used.

The MCS receives and processes data from all SBIRS sensors and manages the SBIRS satellites. MCS software issues launch reports, noting missile type, launch point, time, **azimuth**, and predicted impact point. Finally, U.S. Space Force operators review the launch reports and release them to need-to-know users worldwide.

What's Next?

When it comes to satellite and rocket technology, advancements are constantly being made. By the time you finish reading this book, there may be brand-new technology to replace the older systems. Luckily, whenever space is involved, there are infinite possibilities for exploration, so these advancements will always be welcome.

Join the Debate

1. This book has covered a lot of accomplishments that were achieved because of satellites. Which do you think was the most significant? Why?
2. Do you think getting the first people to the moon was an important accomplishment? Why or why not?
3. What is one area where you think satellites or rockets could still be improved upon?

GLOSSARY

al-Qaeda: A radical Islamist group organized to engage in terrorist activities.

annexation: The act of taking possession of another country or territory.

azimuth: The angular distance along the horizon to a given object, measured clockwise from the north.

espionage: The act of spying.

integral: Very important or needed.

intercontinental ballistic missile: A guided missile with a range of more than 3,300 miles (5,310 km).

oxidizer: An agent that combines oxygen with another element.

propellant: A propelling agent, such as fuel, that provides thrust for a rocket.

reconnaissance: An exploration or examination of an area to gain information.

reentry spacecraft: A returnable spacecraft capable of reentering Earth's atmosphere and being recovered.

resolution: The degree of detail captured by a camera.

telemetry: An automated communications process that collects measurements and other data at some remote point and transmits them to receiving equipment for monitoring, display, and recording.

thrust: A push forward.

Union of Soviet Socialist Republics (USSR): The former union of 15 Soviet republics governed from the Russian capital of Moscow; the union was formally dissolved in 1991.

FIND OUT MORE

Books

Baker, David. *Inside US Spy Satellites*. New York, NY: Rosen Publishing, 2019.

Furgang, Kathy. *Rocket Science and Spacecraft Fundamentals*. New York, NY: Britannica Educational Publishing, 2018.

Klatte, Kathleen A. *Jobs in the U.S. Space Force*. New York, NY: Rosen Publishing, 2023.

Websites

How Satellites Work
science.howstuffworks.com/satellite.htm
Check out this website to learn the ins and outs of satellites.

SpacePlace: Satellites
spaceplace.nasa.gov/search/satellites/
NASA's Space Place website has tons of information about satellites.

What Is a Rocket?
www.nasa.gov/audience/forstudents/k-4/stories/nasa-knows/what-is-a-rocket-k4.html
Read this article from NASA to learn even more about rockets.

Organizations

U.S. Air Force
1670 Air Force Pentagon
Washington, DC 20330-1670
www.airforce.com
Despite no longer managing the operation and study of all satellites and rockets in the military, members of the U.S. Air Force still contribute greatly to this field. Find out more about this branch of the military.

U.S. Space Force
2020 U.S. Space Force Pentagon, STE 4E858
Washington, DC 20330-2000
www.spaceforce.mil
Find out all about the newest branch of the U.S. military, including possible career paths in the force.

Publisher's note to educators and parents: Our editors have carefully reviewed these websites to ensure that they are suitable for students. Many websites change frequently, however, and we cannot guarantee that a site's future contents will continue to meet our high standards of quality and educational value. Be advised that students should be closely supervised whenever they access the internet.

INDEX

ABOUT THE AUTHOR

Tanner Billings is a writer living in New York. He enjoys studying the history of the military and keeping up with modern military technologies. When he is not writing, he enjoys hiking and taking trips to national parks with his family and dogs.